Sample Exams
for
Electrical Licensing

by

Michael G. Owen

Instructions for Examinations

The following tests are typical of examinations given by various electrical authorities to determine the qualifications of individuals seeking their electrical license. However, these examinations cover only the *National Electrical Code®* and electrical theory: there are no questions that pertain to the city ordinance or power company standards of any geographical area. Prior to testing in any location, one should obtain these various documents to insure compliance therewith.

These examinations are divided into two major categories; "CLOSED-BOOK" and "OPEN-BOOK".

A "CLOSED-BOOK" examination is intended to test one's recall of various regulations of the *National Electrical Code®*. These test are to be taken using absolutely no reference materials. A calculator may be used when necessary.

An "OPEN-BOOK" exam is designed to test the individual's ability to use the *National Electrical Code®* in making necessary calculations and other determinations. No reference material, other than the *National Electrical Code®*, Vol. 70, should be used when taking these examinations. A calculator may be used when necessary.

It is imperative that a quite, clean and comfortable setting be maintained during these and any examinations; total concentration is necessary. Do not allow distractions or interruptions during the time allotted for exams.

Time allotments:	Grading/points per question:
Test #1..............1.5 hrs	.2 points
Test #2.............. 2.5 hrs	.5 points
Test #3.............. 2.5 hrs	.2 points
Test #4.............. 2.5 hrs	.2.7 points
Test #5.............. 1.5 hrs	.2.9 points
Test #6.............. 1.5 hrs	.4.4 points
Test #7.............. 2 hrs	.5.6 points

Notes:
Unless otherwise specifically stated, assume the following conditions to exist:

1. All conductor terminals to be rated for 75ºC
2. No more than three conductors in any raceway and 86ºF ambient temperature
3. $\sqrt{3} = 1.732$

Electrical Examination
Test #1

This examination is designed to test the individual's knowledge of the *National Electrical Code®*. There are no questions relating to any city ordinance or power company electrical standards. Circle the answer that **most** appropriately applies to the question. Do not circle more than one answer; doing so will result in no credit being given for that particular question. Time allotment... 1.5 hours, 2 points per question.

This is a "CLOSED-BOOK" examination. No reference material to be used.

1. In the bedroom of a dwelling unit, receptacle outlets shall be installed so that no point in any wall space is further than ____ feet from an outlet in that wall space.

A. 6   B. 8   C. 10   D. 12   E. 15

2. The small appliance branch circuit in a one-family dwelling is allowed to serve an outdoor receptacle.

A. True   B. False

3. An outlet installed in the garage of a dwelling that is to serve the laundry equipment is required to be protected by a ground-fault protective device.

A. True   B. False

4. The refrigerator in the kitchen of a dwelling unit is required to be served by an individual branch circuit.

A. True   B. False

5. A receptacle outlet shall be required at each counter space ____ inches or more in width in all kitchens of dwelling units.

A. 10   B. 12   C. 18   D. 24   E. 30

Sample Exams for Electrical Licensing by Michael G Owen, Test #1
All rights protected by copyright.

6. For a one-family dwelling at least one receptacle outlet in addition to any provided for the laundry equipment shall be installed in each basement and in each _____ garage.

A. single   B. double   C. attached   D. separate   E. grade-level

7. For household ranges of _____ KW or more rating, the minimum branch-circuit rating shall be 40 amperes.

A. 8   B. 8 3/4   C. 10   D. 12   E. 12 3/4

8. A 40 gallon water heater has a nameplate rating of 4.5 Kw, 240 volts, 1ø. What is the minimum standard rating of the branch circuit required to serve this load ?

A. 23.4 amperes   B. 18.75 amperes   C. 30 amperes   D. 25 amperes

9. When using a metal underground water pipe that is in direct contact with the earth for 10' or more as the grounding electrode, is it necessary to supplement this electrode with another grounding electrode, such as a ground rod ?

A. No   B. yes

10. Service drops of 120/240v, 1ø, 3 wire crossing a residential driveway must have a clearance of _____ feet above the driveway.

A. 10   B. 12   C. 15   D. 18   E. 16

11. Recessed incandescent fixtures shall have _____ _____.

A. switched lampholders
B. solid lenses
C. thermal protection
D. screw-shell lampholders

12. Thermal insulation shall not be installed within ___ inches of the recessed fixture enclosure, wiring compartment, or ballast, and shall not be so installed above the fixture so as to entrap heat and prevent the free circulation of air.

A. 1/2   B. 2   C. 3   D. 4   E. 6

13. Fixtures supported by the framing members of a suspended type ceiling are required to be bolted or other wise fastened to the framing members.

A. True   B. False

14. A fixture that weighs more than _____ pounds shall be supported independently of the outlet box.

A. 16   B. 10   C. 24   D. 35   E. 50

15. Fixture wires shall not be used as _____ _____ .

A. Permanent wiring   B. control conductors   C. branch-circuit conductors

16. When lighting track is installed in a continuous row, each individual section of not more than _____ feet in length shall have one additional support.

A. 3   B. 4   C. 5   D. 6   E. 8

17. A receptacle installed outdoors where exposed to the weather or in other wet locations shall be in a _____ enclosure, the integrity of which is not affected when the receptacle is in use.

A. raintight   B. rainproof   C. weathertight   D. weatherproof   E. metal

18. The enclosure for a receptacle installed in an outlet box flush mounted on a wall surface shall be made weatherproof by means of a weatherproof faceplate assembly that provides a _____ connection between the plate and the wall surface.

A. waterproof   B. watertight   C. weatherproof   D. raintight

19. A _____ ampere rated overcurrent device is the smallest standard rated permitted for a 15 KW, 240 volt, 1ø, fixed electric space heater.

A. 60   B. 65   C. 70   D. 80   E. 100

20. Conductors of type NM cable shall be rated at _____ ºC. The ampacity of type NM cable shall be that of _____ ºC conductors in Table 310-16.

A. 60...60   B. 90....90   C. 90....60   D. 60....90   E. 75.....75

21. The ampacity of #8/3 copper type NM cable, where the cable is extended through an attic whose temperature is 110°F is _____ amperes.

A. 40   B. 45   C. 47.85   D. 34.6   E. 34

22. Type NM cable shall be secured in place at intervals of _____ feet and within ____ inches from each cabinet, box or fitting.

A. 4....12   B. 3.5....12   C. 4.5.....12   D. 4.5....8   E. 3.5....8

23. Where holes are bored through wood framing members for the installation of cable type wiring methods, the holes shall be bored so that the hole is not less than _____ inches from the nearest edge of the framing member.

A. 1-1/4   B. 1-1/2   C. 1-3/4   D. 1   E. 2

24. An underground residential 120 volt, 20 ampere branch circuit is allowed a minimum cover of six inches, where extended through an area not subjected to physical damage.

A. True   B. False

25. The direct-burial cable used as service lateral conductors serving a 120/240 volt, 1ø, 3 wire residential service are required to have a minimum cover of _____ inches below finished grade.

A. 12   B. 18   C. 24   D. 30   E. 36

26. When installing a conduit, the maximum degrees of bends can not exceed _____ between boxes or fittings.

A. 180   B. 270   C. 260   D. 360   E. 90

27. Where installing 3" rigid metal conduit in a straight run, using threaded couplings, it is permissible to support the conduit at intervals not to exceed _____ feet.

A. 10   B. 12   C. 15   D. 20   E. 25

28. The maximum distance between supports for 1" rigid nonmetallic conduit shall be _____ feet.

A. 3   B. 5   C. 7   D. 10   E. 12

29. It is permissible to use 3/4" rigid nonmetallic conduit for the support of fixtures, where the fixtures do not exceed 12" above finished grade.

A. True   B. False

30. The maximum number of 1.66 ampere rated store lighting fixtures that may be supplied by a 20 ampere branch circuit is _____.

A. 8   B. 9   C. 10   D. 11   E. 12

31. It is permissible to install a 15 ampere rated single receptacle on an individual branch circuit with a rating of 20 amperes.

A. True   B. False

32. "35" amperes is a standard rating for inverse-time circuit breakers.

A. True   B. False

33. The smallest standard rated fuse is one ampere.

A. True   B. False

34. Where installed in raceways, conductors of size #_____ and larger shall be stranded.

A. 10   B. 8   C. 6   D. 4   E. 3

35. A lighting fixture in a permanently installed pool shall be installed with the top of the fixture lens at least _____ inches below the normal water level of the pool.

A. 12   B. 16   C. 18   D. 20   E. 24

36. Class ___ locations are those hazardous because of the presence of combustible dust.

A. I   B. II   C. III   D. IV   E. V

37. Continuous load: A load where the _____ current is expected to continue for three hours or more.

A. flow-of   B. continuous   C. maximum   D. circuit's   E. load

38. The maximum number of #12 THWN conductors that may be installed in a 1/2" rigid metal conduit is _____ .

A. 3   B. 5   C. 7   D. 9   E. 12

39. When installing six current carrying conductors in the same raceway, the ampacity of those conductors as given in table 310-16 shall be reduced to ____%.

A. 80   B. 70   C. 60   D. 50   E. not reduced

40. The rating of a branch circuit supplying a continuous load shall not be less than _____% of the continuous load.

A. 125   B. 75   C. 100   D. 75   E. 80

41. Receptacles on the property shall be located at least _____ feet from the inside wall of the pool.

A. 5   B. 7   C. 10   D. 15   E. 20

42. At least one receptacle outlet shall be installed directly above a show window for each ____ linear feet or major fraction thereof of show window measured horizontally at it's maximum width.

A. 12   B. 10   C. 21   D. 16   E. 18

43. In multiwire circuits, the continuity of a _____ conductor shall not depend upon device connections, such as lampholders, receptacles, etc., where removal of such device would interrupt the continuity.

A. grounded   B. ungrounded   C. energized   D. phase

44. Conductors which are intended for use as ungrounded conductors, whether used as single conductors or in multiconductor cables, shall be finished to be clearly distinguishable from _____ conductors.

A. grounded   B. other   C. communication   D. branch-circuit

45. Where used outside, aluminum or copper-clad aluminum grounding conductors shall not be installed within ____ inches of the earth.

A. 12   B. 18   C. 24   D. 30   E. 36

46. The minimum number of 20 ampere, 120 volt, general illumination circuits required for a dwelling unit with 2500 square feet of area, exclusive of open porches, garages and areas not adaptable for future use, is _____ .

A. 3   B. 4   C. 5   D. 6   E. 8

47. A _____ conductor shall not be counted when applying the provisions of Note 8 to Tables 310-16 or 310-18.

A. deenergized   B. fiber optic   C. grounded   D. service   E. grounding

48. Single conductors specified in Table 310-13 shall only be permitted to be installed where part of a recognized wiring method of Chapter _____ .

A. 1   B. 2   C. 3   D. 4   E. 5

49. At least ___ inches of free conductor must be left at each outlet box for the connection of devices and for splicing (not mobile homes).

A. 4   B. 6   C. 8   D. 10   E. 12

50. Receptacles that provide shore-power for boats over 20' in length shall be of the _____ and grounding type and rated at 30 amperes or more.

A. locking   B. 3-wire   C. flush   D. duplex   E. weatherproof

End of Test #1

Electrical Examination .... Test #2

This examination is designed to test the individual's ability to determine the size or ratings of the conductor and overcurrent devices required to serve various types of electrical equipment. Base all answers on the *National Electrical Code* ®. The practical use of electrical theory may be necessary to determine the proper requested values. All formulas and /or calculations must be shown in order to receive credit for the individual problems. Circle the correct or most correct answer. Do not circle more than one answer; doing so will result in no credit being issued for that particular problem/question. Time allotment, 1.5 hours; 5 points per question.

This is an "OPEN-BOOK" Examination. Any reference material may be used.

1. A 20 ampere, 1ø load is located at a distance of 183 feet from it's source, where the voltage actually measures 118. Using #10 copper conductors to serve this load, determine the voltage that is available at the load. Use "12" as the constant for copper.

A. 8.5   B. 109.5   C. 112.7   D. 113.8   E. 114.2

2. Determine the size wire, in AWG, required to serve a 40 ampere, 1ø load at a distance of 154 feet from it's source, where the voltage is actually 118. Use copper conductors with a constant of 12. Size conductors for a 3% voltage drop.

A. 8   B. 6   C. 4   D. 3   E. 2

3. A 3ø, 75 ampere load is located 272.18 feet from it's source where the voltage is 480. Determine the circular mil area of the aluminum conductors required to serve this load maintaining a 3% voltage drop. Use "17" as the constant for aluminum.

A. 16,510   B. 6530   C. 52,620   D. 41740   E. 26,240

4. Determine the minimum standard rating of the overcurrent device required to serve a continuous load of 8.16 KVA at 120 volts, 1ø.

A. 90   B. 100   C. 65   D. 85   E. 70

5. A 3ø, 240 volt non-continuous load of 24.7 KW operates with a power factor of 70%. Determine the minimum ampere rating of the conductors required to serve this load.

A. 85   B. 100   C. 65   D. 68   E. 70

6. A conduit containing 6#1/0 THWN copper current carrying conductors passes through an area with an ambient temperature of 106ºF. Determine the ampacity of each individual conductor under this condition of use.

A. 150   B. 123   C. 137.6   D. 98.4   E. 128.3

7. Determine the maximum standard rating of the inverse-time circuit breaker allowed to serve as the overcurrent device for a 240 volt, 1ø, 5.6 KW, 50 gallon storage type water heater.

A. 25   B. 30   C. 35   D. 40   E. 45

8. Determine the maximum standard rated time-delay fuse that may protect the primary of a 25 KVA, 1ø transformer having a primary voltage of 480 volts.

A. 100   B. 110   C. 125   D. 150   E. 175

9. Determine the maximum KVA of load that may be installed on two-25 KVA transformers that are connected in "open-delta" providing a 120/240 volt, 3ø secondary voltage without creating an overload condition.

A. 50   B. 43   C. 60   D. 70   E. 80

10. Determine the minimum trade size rigid metal conduit required to contain 3#4THWN, 3# 1THWN and 1#6 bare conductors.

A. 3/4"   B. 1"   C. 1-1/4"   D. 1-1/2"   E. 2"

11. Determine the maximum number of #4/0 THWN conductors that may be installed in a wireway that has interior measurements of 4"x4".

A. 8   B. 10   C. 12   D. 14   E. 16

12. In the living room of a dwelling unit, a particular wall space measures 24 feet. Determine the minimum number of receptacle outlets that are required in this area.

A. 0   B. 1   C. 2   D. 3   E. 4

13. In a commercial office area, a particular wall space measures 24 feet. Determine the minimum number of receptacle outlets that are required in this area.

A. 0   B. 1   C. 2   D. 3   E. 4

14. An office building has outside dimensions of 85'-3" x 95'-9" and is to be served by a 120/240 volt, 1ø service. Determine the minimum number of 120 volt, 20 ampere branch circuits required to serve the general illumination load.

A. 10   B. 11   C. 12   D. 13   E. 15

15. Determine the minimum AWG rating of the conductors serving a 240 volt, 1ø, 3 Hp swimming pool pump motor that has a continuous rating. The motor will operate for 12 consecutive hours before cycling off. The conductors are to be installed in a conduit containing no other conductors.

A. #12 THWN   B. #14 THWN   C. #10 THWN   D. #10 TW Aluminum

16. Calculate the minimum cubic-inch volume of the nonmetallic box required to contain 3 #12/2 AWG NM cable and one duplex receptacle. The box is to contain no other devices or equipment.

A. 14   B. 16   C. 20.25   D. 22.25   E. 24.5

17. Determine the minimum depth of a metal 3"x2" device box required to contain one duplex receptacle, 2#12/2 AWG NM cables and two cable clamps.

A. 2   B. 2-1/4   C. 2-1/2   D. 2-3/4   E. 3-1/2

18. Determine the minimum size copper grounded conductor that may be used in the service where the largest service-entrance conductors are #500 kcmil copper.

A. #6   B. #4   C. #2   D. #1/0   E. #2/0

19. A conduit is extended between two switch enclosures in a Class I- Division 1 location. The total length of the conduit measures 36". Determine the minimum number of seals that are required for this conduit run.

A. 1   B. 2   C. 3

20. The terminal bar shall be bonded to the cabinet or panelboard frame, if of metal, otherwise it shall be connected to the _____ _____ that is run with the conductors feeding the panelboard.

A. grounding conductor   B. grounding electrode   C. grounded conductor

End of Test 2

Electrical Examination
Test #3

Each of the following questions are to be answered **TRUE** or **FALSE**. Circle "T" for true and circle "F" for false. Provide the *National Electrical Code* ® section number that best supports your answer. Failure to provide this reference will result in no credit being given for that particular question. If any portion of the statement is incorrect or false, consider the entire statement false. Allotted time, 2 hours; 2 points per question.

This is an "OPEN-BOOK" examination.

1. _____ T F All 125v, 15 and 20 ampere receptacles installed in all bathrooms must be protected by a ground-fault circuit-interrupter.

2. _____ T F An outdoor receptacle may be supplied by a small appliance branch circuit of a dwelling unit.

3. _____ T F A lighting fixture installed in a clothes closet of a dwelling unit may be controlled by means of a "pull-chain".

4. _____ T F It is permissible to install electrical metallic tubing in direct contact with the earth.

5. _____ T F An autotransformer may not be used to supply branch circuits.

6. _____ T F Installations in compliance with the *National Electrical Code*® will provide a safe and adequate service.

7. _____ T F For equipment rated 1200 amperes or more and over 6 feet wide, ... , there shall be one entrance not less than 24" and 6-1/2 feet high at each end.

8. _____ T F Time switches, flashers, and similar devices need not be of the externally operable type.

9. _____ T F A Class III, Division 1 location is a location in which easily ignitable fibers are stored or handled.

10. _____ T F Control circuit devices with screw-type pressure terminals used with #14 AWG or smaller copper conductors shall be torqued to a minimum of 7 pound-inches unless identified for a different torque value.

11. _____ T F Auxiliary gutters shall not contain more than thirteen current carrying conductors at any cross-section.

12. _____ T F The maximum size of liquidtight flexible nonmetallic conduit shall be the 3 inch trade size.

13. _____ T F Listed ceiling fans that do not exceed 50 pounds in weight, with or without accessories, shall be permitted to be supported by outlet boxes identified for such use...

14. _____ T F The word transformer is intended to mean an individual transformer, not a bank of individual transformers.

15. _____ T F The minimum size traveling cable for controls to an elevator is #14 AWG.

16. _____ T F A grounding or bonding conductor shall not be counted when applying the provisions of Note 8 the Ampacity Tables.

17. _____ T F Spacing between conduits, tubing, or raceways shall be maintained.

18. _____ T F Where it is unlikely that two dissimilar load will operate simultaneously, it shall be permissible to omit the lesser of the two loads.

19. _____ T F For devices with screw shells, the terminal for the grounded conductor shall be the one connected to the screw shell.

20. _____ T F The bottom of a sign shall not be less than 18 feet above areas accessible to vehicles.

Sample Exams for Electrical Licensing, Michael G. Owen, Test #3
All rights protected by copyright.

21. _____ T   F   Special permission is the written consent of the authority having jurisdiction.

22. _____ T   F   Wall switches shall be located at least 10 feet from the inside walls of a hot tub.

23. _____ T   F   Class 1 circuits and equipment shall not be required to be grounded in accordance with Article 250.

24. _____ T   F   Electrically powered equipment whose continuous operation is necessary to maintain a patients life is known as "Emergency Life Support Equipment".

25. _____ T   F   A thermal barrier shall be required if the space between the resistors and the reactors and any combustible material is less than 12 inches.

26. _____ T   F   Nominal battery voltage shall be 1.2 volts per cell for the lead-acid type battery.

27. _____ T   F   Type THW insulated conductor may be used in dry or wet locations.

28. _____ T   F   Fixture wire shall be permitted to be used as general purpose branch circuit wiring if protected from physical damage.

29. _____ T   F   Where the service raceway extends through the roof and the service drop cable is attached to the raceway, the *NEC®* requires that the conduit be a minimum of 2 inch trade size rigid metal conduit.

30. _____ T   F   The overall length of the supply cord for a mobile home shall not be less than 21 feet.

31. _____ T   F   The maximum size of electrical nonmetallic tubing shall be the 1-1/4 inch trade size.

32. _____ T   F   It is permissible to use Type NM cable in exposed work.

33. _____ T F A bathroom is an area containing a tub or a shower.

34. _____ T F The ampacity correction factor for Type THWN insulation in an area with an ambient temperature of 115°F is .75.

35. _____ T F Direct Buried Cables, 600v or less, shall have a minimum cover of 30 inches.

36. _____ T F Overcurrent devices shall not be located in the vicinity of easily ignitable fibers such as in clothes closets.

37. _____ T F Plug fuses shall be permitted for use in circuits exceeding 150v to ground.

38. _____ T F A standard classification for a multioutlet branch circuit is 25 amperes.

39. _____ T F Wire mesh provided in the concrete floor of animal confinement areas to provide an equipotential plane shall be bonded to the building grounding electrode system.

40. _____ T F Placing a knot in a flexible cord is an acceptable method that may be used to relieve the tension on terminals.

41. _____ T F The ampacity correction factors of Note 8 to the Ampacity Tables do not apply to conductors installed in a nipple that does not exceed 24 inches in length.

42. _____ T F In a completed seal, the thickness of the sealing compound shall not be less than the trade size of the conduit and shall not be less than 1/2 inch.

43. _____ T F The ampacity of the phase conductors from the generator terminals to the first overcurrent device shall not be less than 125% of the nameplate current rating of the generator.

44. _____ T F Where a flexible cord is used to supply a 240v room air-conditioner, the length of the cord shall not exceed 4 feet.

45. _____ T F The impedance heating elements shall not operate at a voltage greater than 24 volts, ac.

Page # 4

Sample Exams for Electrical Licensing, Michael G. Owen, Test #3
All rights protected by copyright.

46. _____ T F The full-load current rating for a 10 Hp, 208v, 3ø motor is 30.8 amperes.

47. _____ T F The DC resistance of #1 uncoated copper conductor at 75°C is .154Ω per 1000 feet.

48. _____ T F Where practical, a separation of at least 10 feet shall be maintained between any coaxial cable and lightning conductors.

49. _____ T F Bonding shall be provided where necessary to assure electrical continuity and the capacity to conduct safely any fault current likely to be imposed.

50. _____ T F Snap switches rated 30 amperes or less directly connected to aluminum conductors shall be listed and marked CO/ALR.

End of Test #3

Electrical Examination
Test #4

The following questions or problems are of the multiple choice type. Circle the answer that most accurately applies. Do not circle more than one answer per question; doing so will result in no credit being issued for that particular question or problem. Provide the *National Electrical Code®* section that best supports your answer. Failure to support your answer will result in a reduction of 1/2 of the credit for that particular question or problem. Time allotment, 2 hours; 2.7 points per question.

This is an "OPEN-BOOK" examination.

1. The outlet for a sign for a commercial building is required to be served by a ....

A. 120v multiwire branch circuit
B. 20 ampere branch circuit
C. weatherproof duplex receptacle outlet
D. 1200va branch circuit

Section_____

2. Type XHHW insulated conductors may be used in
A. Dry locations only
B. Wet locations only
C. Dry, damp or wet locations
D. None of the above locations

Section_____

3. The full load current for a 1/6 Hp, 115 volt, 1ø AC motor is _____ .

A. 4.4   B. 4.8   C. 6   D. 4

Table_____

4. A 3"x2"x2" device box can hold _____ #12 conductors, as long as there are no other devices or fittings within the box.

A. 3   B. 4   C. 5   D. 7

Table_____

Page # 1

Sample Exams for Electrical Licensing, Michael G. Owen, Test #4
All rights protected by copyright

5. The conductors that supply a 185 ampere (primary) transformer welder with a duty cycle of 60% must have a minimum ampacity of _____ .

A. 111	B. 144	C. 185	D. 231.25

Section_____

6. Where the rated primary current of the transformer is less than 9 amperes, an overcurrent device rated or set at not more than ____% of the primary current shall be permitted.

A. 125	B. 137	C. 150	D. 167

Section_____

7. A 4.8kw, 240v, 1ø, 40 gallon water heater must be served by a branch circuit that has a rating not less than _____ .

A. 20	B. 25	C. 30	D. 35

Section_____

8. Where conduits enter the bottom of a floor-standing panelboard, the conduits shall not extend more than ____ inches above the bottom of the enclosure.

A. 2	B. 3	C. 4	D. 6

Section_____

9. What is the smallest AWG size fixture wire that may be connected to a 120v, 50 ampere branch circuit.

A. 12	B. 10	C. 8	D. 14

Section_____

10. The professional projector is a type using 35 or 70 mm film with a minimum width of 1-3/8", with _____ perforations per inch.

A. 5.4	B. 4.5	C. 6.8	D. 8.6

Section_____

11. Direct buried conductors or cables shall _____ permitted to be spliced without the use of a splice box.

A. be   B. not be

Section_____

12. The ampacity of a #4 THWN copper conductor, installed in a raceway with three other current-carrying conductors, in an area with an ambient temperature of 97°F is _____ .

A. 85   B. 68   C. 59.84   D. 52.36

Table _____

13. The maximum operating temperature of Type NM cable shall be ____°C.

A. 60   B. 75   D. 85   D. 90

Section _____

14. The maximum operating temperature of Type PTF conductors is ____°C.

A. 60   B. 90   C. 200   D. 250

Table _____

15. The maximum number of #12 THWN conductors that may be installed in a 2" E.M.T. nipple that is 24" in length is _____ .

A. 114   B. 151   C. 164   D. 70

References:_____ .

16. Optical fiber cables can be grouped into _____ types.

A. 2   B. 3   C. 4   D. 5

Section _____

17. An insulated grounded conductor of No. ____ or smaller shall be identified by a continuous white or natural grey outer finish along its entire length.
A. 8   B. 6   C. 4   D. 2

Section_____

18. The minimum elevation of 7.2kv unguarded live part above a working area is _____ .

A. 8'	B. 8'6"	C. 9'6"	D. 10'

Table _____

19. A #12 AWG conductor has a circular mill area of _____ cm.

A. 6350	B. 6850	C. 6530	D. 3650

Table _____

20. Underground wiring shall not be permitted under the pool or under the area extending ____ feet horizontally from the inside wall of the pool.

A. 5	B. 10	C. 15	D. 20

Section _____

21. Electrical nonmetallic tubing shall be secured at least every ___ feet.

A. 3	B. 5	C. 8	D. 10

Section _____

22. The switch or switches installed in emergency lighting circuits shall be so arranged that only _____ persons will have control of the emergency lighting.

A. qualified	B. management	C. security	D. authorized

Section _____

23. What is the ampacity of #10 THW copper conductors when installed in a raceway with 3 other conductors and serves a 30-minute rated hoist ?

A. 43	B. 30	C. 35	D. 48

Table _____

24. The calculated load for a household type range rated at 12.6 KW is _____ va.

A. 8000	B. 8400	C. 8500	D. 12600

Table_____

25. When Type THWN insulated conductors are installed in a 4" rigid metal conduit, the minimum radius of a 90º bend in the conduit must not be less than ____ inches.

A. 18	B. 21	C. 24	D. 27

Table _____

26. The diameter of #500 kcmil THW is _____ inches.

A. .8316	B. .0005	C. 1.003	D. .5000

Table _____

27. Where a extra-hard usage flexible cord is used to cord and plug connect a free-standing type partition, the maximum length of the cord shall be _____ feet.

A. 2	B. 3	C. 4	D. 6

Section _____

28. Where ungrounded conductors No. ____ or larger enter a raceway in a cabinet, box or gutter, the conductors shall be protected by a substantial fitting providing a smoothly rounded insulating surface.

A. 6	B. 4	C. 2	D. 1/0

Section _____

29. Direct buried cables _____ permitted to be spliced or tapped without the use of splice boxes.

A. shall be	B. shall not be

Section _____

30. What is the maximum cord and plug connected load that may be connected to a 15 ampere, 120 volt receptacle that is installed on a 20 ampere, multioutlet branch circuit?

A. 15 amperes   B. 20 amperes   C. 12 amperes   D. 16 amperes

Table _____

31. A commercial occupancy has 150, 15 ampere, 120 volt receptacles installed for general use. A load of not less than _____volt-amperes must be added to the service to compensate for these receptacles.

A. 180   B. 13,500   C. 18,500   D. 27,000

Section _____ Table _____

32. When #1/0 conductors are installed in parallel, they must be of the same conductor material.

A. True   B. False

Section _____

33. The ampacity of #8/3c Type SO cord is _____ amperes.

A. 40   B. 35   C. 30   D. 50

Table _____

34. Aluminum cable trays shall not be used as equipment grounding conductors for circuits with ground-fault protection above _____ amperes.

A. 600   B. 800   C. 1000   D. 2000

Table _____

35. A wall space in the den of a dwelling measures 24 feet in length, unbroken along the floor line. The minimum number of receptacles that are required to be installed in this wall space is _____ .

A. 1   B. 2   C. 3   D. 4

Section _____

36. The minimum AWG size of an aluminum equipment grounding conductor for a circuit being protected by a 800 ampere overcurrent device is _____ .

A. 1/0   B. 1   C. 3/0   D. 3

Table _____

37. The maximum operating temperature of Type THWN insulated conductors is _____ °F.

A. 30   B. 86   C. 90   D. 167

Table _____

End of Test #4

Electrical Examination
Test #5

This examination is designed to test the individual's knowledge of the *National Electrical Code*®. Circle the one answer that most appropriately applies to the question. Do not circle more than one answer; doing so will result in no credit being given for that particular question. Time allotment, 1.5 hours; 2.9 points per question.

This is a "CLOSED-BOOK" examination.

1. What is the minimum size TW copper conductor that may be used as branch circuit conductors serving a 40 gallon, 240v, single phase, 6 KW water heater ?

    A. #12   B. #10   C. # 8   D. # 6

2. Overhead conductors for festoon lighting shall not be smaller than # __.

    A. #14   B. #12   C. #10   D. # 8

3. Two boxes are mounted on the same wall and are 66' apart. There is to be a 3/4" rigid metal conduit extended between the two boxes. Determine the minimum number of straps that are required to support this run of conduit.

    A. 4   B. 5   C. 6   D. 7

4. What is the minimum burial depth for rigid metal conduit?

    A. 6"   B. 12"   C. 18"   D. 24"

5. Define the following terms:

    A. Bathroom:

    B. Readily accessible:

    C. Dusttight:

    D. Grounded conductor:

6. What is the maximum number of #12 THWN conductors that may be installed in a 1/2" rigid metal conduit?

    A. 3   B. 5   C. 7   D. 9

7. Give the minimum lighting load per square foot of the following occupancies:
    A. Churches: _____va/sq.ft.
    B. Schools: _____va/sq.ft.
    C. Stores: _____va/sq.ft.
    D. Banks: _____va/sq.ft.

8. A load of not less than _____volt amperes per linear foot of show window shall be permitted instead of the specified unit load per outlet.

    A. 1.5   B. 180   C. 200   D. 500

9. A _____ conductor shall not be counted when applying the provisions of Note 8 to the Ampacity Tables.

    A. deenergized   B. fiber-optic   C. grounded   D. grounding

10. When installing twenty conductors in the same raceway, the ampacities given in Table 310-16 shall be reduced to _____%.

    A. 80%   B. 70%   C. 60%   D. 50%

11. The ampacity of the phase conductors from the generator terminals to the first overcurrent device shall not be less than _____% of the nameplate current rating of the generator.

    A. 115%   B. 125%   C. 135%   D. 150%

12. Class I locations are those that are hazardous because of the presence of _____.

    A. Gases or vapors   B. Combustible dust
    C. Fibers or flyings   D. Petroleum products

13. Underground wiring shall not be permitted under the pool (swimming) or under the area extending _____feet horizontally from the inside wall of the pool.

    A. 5   B. 7   C. 10   D. 15

14. Class 1 Power-limited circuits shall be supplied from a source having a rated output of not more than ____volts and ____volt-amperes.

    A. 125...1200   B. 30...1000   C. 50...1000   D. 24...1200

15. Give the article number for the article entitled "Electric Welders".

    A. 250   B. 440   C. 630   D. 680

16. How many receptacle outlets are required above a show window that measures 60' at it's maximum width.

    A. 4   B. 5   C. 6   D. 7

17. The total load on any overcurrent device in a panelboard shall not exceed ____% of its rating where in normal operation the load will continue for 3 hours or more.

    A. 125%   B. 75%   C. 80%   D. 100%

18. Conductor sizes are expressed in _____ _____ _____ (AWG) or in circular mils.

19. Define the following terms:

    A. Automatic:

    B. Ampacity:

    C. Dwelling unit:

    D. Festoon lighting:

20. The minimum headroom of working spaces about service equipment, switchboards, panelboards, or motor control centers shall be ____feet. (600v or less)

    A. 6   B. 6.25   C. 6.5   D. 8

21. For devices with screw shells, the terminal for the _____ conductor shall be the one connected to the screw shell.

    A. switched   B. ungrounded   C. grounded   D. grounding

22. A heavy-duty lampholder shall have a rating of not less than ____watts if of the admedium type and not less than ____watts if of any other type.

   A. 600....750   B. 660....750   C. 660....700   D. 600....700

23. Give the minimum lighting load for the following type occupancies:

   A. Hospitals   = _____va/sq.ft.
   B. Restaurants = _____va/sq.ft.
   C. Barber shop = _____va/sq.ft.

24. Receptacles on the property shall be located at least ____feet from the inside walls of a pool.

   A. 5   B. 10   C. 15   D. 20

25. Receptacles on the property shall be at least ____ feet, measured horizontally, from the inside walls of a spa or hot tub.

   A. 5   B. 10   C. 15   D. 20

26. Temporary electrical power and lighting installations shall be permitted for a period not to exceed ____ days for Christmas decorative lighting, carnivals, and similar purposes.

   A. 180   B. 120   C. 100   D. 90

27. Plug fuses of the Edison-base type shall be classified at not over ____volts and ____amperes and _____ .

   A. 250....1....above   B. 125....30....below
   C. 150....1....below   D. 250......0...above

28. The screw shall of a plug fuse shall be connected to the ____side of the circuit.

   A. Supply   B. Load   C. Secondary   D. Primary

29. For ranges of 8 3/4 KW or more rating, the minimum branch circuit rating shall be _____amperes.

   A. 35   B. 40   C. 45

30. An autotransformer is a transformer in which a part of the winding is common to both _____ and _____ circuits.

   A. Feeder....branch
   C. Inductive..resistive
   B. Utility....customer
   D. Primary...secondary

31. At least one receptacle outlet shall be installed directly above a show window for each _____ linear feet or major fraction thereof of show window area measured horizontally at it's maximum width.

   A. 10   B. 21   C. 12   D. 16

32. Single-throw knife switches shall be so connected that the blades are _____ when the switch is in the open position.

   A. Dead   B. Deenergized   C. Insulated   D. Grounded

33. Receptacles that supply shore-power for boats 20' or less in length, shall be rated not less than 20 amperes and shall be of the _____ and _____ types.

   A. duplex.........grounding
   C. single..........grounded
   B. locking.......grounding
   D. grounded....locking

34. Single conductors specified in Tables 310-13 shall only be permitted to be installed where part of a recognized wiring method of Chapter ____.

   A. 2   B. 3   C. 4   D. 6

End of Test #5

Electrical Examination
Test #6

This examination is designed to test the individual's knowledge of the *National Electrical Code®*. Either circle the correct answer or provide a complete answer as required. Time allotment, 1.5 hours, 4.4 points per question.

This is a "CLOSED-BOOK" examination.

1. Define the following terms:

    A. Raceway:

    B. Circuit breaker:

    C. Thermal cutout:

2. The power supply to the mobile home hall be a feeder assembly consisting of not more than _____ tested _____-ampere mobile home supply cord ....

    A. 1.......50   B. 2.......50   C. 1.......60   D. 1......100

3. (**Mobile homes**) Connections of ranges and clothes dryers with 120/240v, 3-wire ratings shall be made with ____-conductor cord and 3 pole, ____wire, grounding-type plugs, or by Type AC cable or conductors enclosed in flexible metal conduit.

    A. 3....3   B. 3....4   C. 4....4   D. 4....3

4. (**Swimming pools**) Conductors on the load side of a ground-fault circuit-interrupter or a transformer shall not occupy conduit, boxes, or enclosures containing _____ conductors.

    A. grounding   B. other   C. bare   D. bonding

5. Pool associated motors shall be connected to an equipment grounding conductor sized in accordance with Article 250-___ but not smaller than #_____. It shall be an insulated copper conductor and shall be installed with the circuit conductors in rigid metal conduit, intermediate metal conduit or rigid nonmetallic conduit.

    A. 95....12   B. 94....12   C. 95....10   D. 94....10

6. In multiwire circuits the continuity of a _____ conductor shall not depend upon device connections, such as lampholders, receptacles, etc., where the removal of such device would interrupt the _____.

    A. phase......................current   B. phase......................current
    C. grounded................continuity   D. grounding...............continuity

7. Metal or nonmetallic raceways, cable armors, and cable sheaths shall be _____ between cabinets, boxes, fittings, or other enclosures or outlets.

    A. continuous   B. supported   C. strapped   D. complete

8. Conductors which are intended for use as _____ conductors, whether used as single conductors or in multiconductor cables, shall be _____ to be clearly distinguishable from grounded and grounding conductors.

    A. phase................insulated   B. energized........colored
    C. grounded..........insulated   D. ungrounded.....finished

9. Optical fiber cables transmit light for control, _____ and _____ through an optical fiber.

    A. signaling..........communications   B. power.................detection
    C. stimulation......detection   D. measurement...power

10. The load for the required branch circuit installed for the supply of exterior signs or outline lighting shall be computed at a minimum of _____ volt-amperes.

    A. 1000   B. 1200   C. 1500   D. 1800

11. The bottom of sign and outline lighting enclosures shall not be less than _____ feet above areas accessible to vehicles.

    A. 10   B. 12   C. 15   D. 14

12. Define the following terms:
    A. Grounded conductor:

    B. Nominal voltage:

    C. Watertight:

13. A pull box is installed in a straight run of 3" conduit containing # 250 kcmil conductors serving a 480 volt load. The minimum length of this pull box is _____.

   A. 12"   B. 18"   C. 21"   D. 24"

14. A conduit contains 24 current carrying conductors. When applying Note 8 to Table 310-16, the ampacity of these conductors will be _____% of the value as given in these tables.

   A. 80   B. 45   C. 60   D. 50

15. Article _____ is the article relating to "*Overcurrent Protection*".

   A. 230   B. 240   C. 420   D. 430

16. In other than _____-_____ dwellings where laundry facilities are not to be installed or permitted, a laundry receptacle shall not be required.

   A. one-family   B. two-family   C. multi-family

17. Where the load is computed on a volt-amperes-per-square-foot basis, the wiring system up to and including the branch-circuit panelboard(s) shall be provided to serve not less than the calculated load. This load shall be evenly proportioned among _____ branch circuits within the panelboard(s).

   A. individual   B. multiwire   C. multioutlet   D. lighting

18. (**Festoon Lighting**) In spans exceeding _____ feet, the conductors shall be supported by a messenger wire; ....

   A. 10   B. 20   C. 30   D. 40

19. Metal enclosures for service conductors and equipment shall be _____.

   A. protected   B. grounded   C. weatherproof   D. galvanized

20. Where used outside, aluminum or copper-clad aluminum grounding conductors shall not be installed within _____ inches of the earth.

   A. 12   B. 18   C. 24   D. 30

21. Where portions of the _____ raceway system are exposed to a widely different temperatures, as in refrigerating or cold-storage plants, circulation of air from a warmer to a colder section through the raceway shall be _____.

    A. exterior....maintained   B. interior.....maintained
    C. exterior....prevented   D. interior.....prevented

22. Give the ampacity of the following THWN insulated copper conductors that are installed in a raceway in an area having an ambient temperature of 75°C.

    A. #12 = _____ amperes
    B. # 8 = _____ amperes
    C. # 3 = _____ amperes
    D. # 1/0= _____ amperes

23. List five conditions which must be met in order to properly parallel # 1/0 and larger conductors.

    1.

    2.

    3.

    4.

    5.

End of Test #6

Electrical Examination
Test #7

This examination is designed to test not only one's knowledge of the *National Electrical Code* ®, but also their understanding of electrical theory. In order to obtain the correct answer to some of the questions, these two must be utilized. Show all work and formulas use to obtain the answers; failure to display work/formulas will result in a reduction of 50% credit for each individual problem effected. Time allotment, 2 hours, 5.6 points per question.

This is an "OPEN-BOOK" examination.

1. Determine the minimum required rating of the overcurrent device supplying a 75 KVA continuous load operating on a 480v, three phase circuit.

2. Calculate the maximum number of #1/0 THWN conductors that may be installed in a 4"x4" wireway. The demand factors of Note 8 to the Ampacity Tables are to be applied.

3. Determine the minimum size rigid metal conduit required to contain 3#1/0 THWN, 3#2THWN and 10 # 14 THWN conductors.

4. Determine the actual current of a fully loaded 3ø, 480v, 15 HP motor that is operating with an efficiency rating of 81% and a power factor of 73%.

Sample Exams for Electrical Licensing, Michael G. Owen, Test # 7
All rights reserved by copyright.

5. Calculate the size of the conductor required to serve a 7.5 KVA, 240v, single phase load that is located a distance of 210 feet from its source, where the voltage is actually 238. Use copper conductors (with a constant of "12") and allow for a 3% voltage drop.

6. An office building has outside dimensions of 57'-6"x 120'-5". Determine the minimum number of 120v, 20 ampere circuits required to serve the general illumination load. The service to the building is to be 120/208v, three phase, 4 wire.

7. Determine the minimum volume of the box required to contain 6#10's power conductors, 10#12's lighting conductors, and 3#14 equipment grounding conductors.

8. Determine the minimum size rigid nonmetallic conduit required to contain one #3/0 bare copper conductor that is to be used as a grounding electrode conductor.

9. Calculate the minimum trade size diameter of a 24" rigid metal conduit nipple that is to contain 30#12THWN conductors.

10. Determine the secondary FLA of a 75 KVA, 3ø, 12470/240v transformer. The transformer has an impedance of 2% and is supplied by the higher voltage.

11. Six #6 THWN copper conductors are installed in the same conduit and are extended through an area with an ambient temperature of 106 °F. Calculate the ampacity of each #6 conductor. All conductors are current-carrying.

12. What is the FLA of a 75 H.P., 208 volt, 3ø, induction motor?

13. Determine the minimum required ampacity of the conductors required to serve a 15 KVAR, 480 volt, 3ø capacitor bank.

14. A 1500 KVA, 12470/480 volt, 3ø transformer is to be installed in a vault that is cooled by the natural circulation of air. Determine the minimum area, in square feet, of the required ventilation opening(s).

15. Determine the minimum size of the copper grounding electrode conductor required for a service that has a bus service head whose ungrounded conductors are 1/4"x4" copper bus bars.

16. Determine the area in square inches of an aluminum bus bar required to serve a 1500 KVA, 480 volt, 3ø load.

17. A 1-1/2 H.P., 115 volt, 3ø motor is to be supplied by a Type SO flexible cord. This is a continuous rated motor used in a continuous service classification. Determine the minimum size of the SO cord.

18. A multiwire circuit is run in conduit and is to serve 3-120 volt, 14 ampere rated data processing units which are to be considered as non-continuous loads. The circuit conductors originate in a 120/208v, 3ø, 4 wire panelboard and have type THWN insulation. Determine the minimum size AWG of these circuit conductors.

End of Test #7

Answer Sheet

TEST #1
1. A   2. B   3. B   4. B   5. B   6. C   7. B   8. D   9. B   10. B
11. C   12. C   13. A   14. E   15. C   16. B   17. D   18. B   19. D   20. C
21. A   22. C   23. A   24. B   25. C   26. D   27. D   28. A   29. B   30. B
31. B   32. A   33. A   34. B   35. C   36. B   37. C   38. D   39. A   40. A
41. C   42. A   43. A   44. A   45. B   46. B   47. E   48. C   49. B   50. A

TEST #2
1. B   2. D   3. D   4. A   5. A   6. D   7. C   8. D   9. B   10. D
11. B   12. C   13. A   14. E   15. A   16. C   17. E   18. D   19. A   20. A

TEST #3: Note-More than one reference may be available for some questions.

1. 210-8,a,b....TRUE
2. 210-52 b ...FALSE
3. 410-8c ....FALSE
4. 348-1.. True
5. 210-9..True
6. 90-1b..False
7. 110-16c.. True
8. 380-5..True
9. 500-7a &b..False
10. 430-9c.. True
11. 374-5..False
12. 351-24..False
13. 422-18..False
14. 450-2&3..False
15. 620-12a2..False
16. Note 11, Tab 310-16..True
17. Note 8b, Tab 310-16..True
18. 220-21.. True
19. 200-10c.. True
20. 600-9a.. False
21. Art. 100.. True
22. 680-41c.. False
23. 725-20..False
24. 517-2.. False
25. 470-3.. True
26. 480-2..False
27. Tab310-13..True
28. 402-11.. False
29. 230-28.. False
30. 550-5d.. True
31. 331-5b.. False
32. 336-3a.. True
33. Art. 100.. False
34. Tab310-16.. True
35. Tab300-5.. False
36. 240-24d.. True
37. 240-50a.. False
38. 210-3.. False
39. 547-8b.. True
40. 400-10(FPN).. True
41. Note 3, Chap9..True
42. 501-5c3.. False
43. 445-5.. False
44. 440-64.. False
45. 426-31.. False
46. Tab430-150.. True
47. Tab8, Chap.9.. True
48. 820-10f3.. False
49. 250-70.. True
50. 380-14c.. False

Test #4: Note-More than one reference may be available for some questions.

1. B...600-5a
2. C..Tab300-13
3. A.. Tab 430-148
4. B.. Tab 370-16a
5. B.. 630-11a
6. D.. 450-3b1, Ex 1
7. B.. 422-14b
8. B.. 384-10
9. A.. 240-4, Ex 2
10. A.. 540-2
11. A.. 300-5e
12. C.. 310-16, Note 8
13. D.. 336-30
14. D.. Tab 402-3
15. B ...Tables 5 & 4, Chap 9, Note 4 under Table 1
16. B.. 770-4
17. B. 200-6a
18. B.. 110-34e
19. C.. Tab 8, Chap 9
20. A.. 680-10
21. A.. 331-11
22. D.. 700-20
23. A.. 610-14a
24. B.. Tab 220-19
25. C.. 346-10
26. C.. TAb5, Chap 9
27. A.. 605-8a
28. B.. 300-4f
29. A.. 300-5e
30. C.. 210-21b-2

Sample Exams for Electrical Licensing...... Michael G. Owen

TEST #4 ... continued

31. C.. 220-3c-5 and Tab 220-13 32. A.. 310-4
33. B.. Tab 400-5   34. D.. Tab 318-7b2**Note 35. B.. 210-52a
36. C.. Tab 250-95   37. D.. Tab 310-13

Test #5:

1. C	2. B	3. D	4. A	5. Art 100
6. D	7. 1..3..3..3½		8. C	9. D
10. D	11. A	12. A	13. A	14. B
15. C	16. B	17. C	18. American Wire Gauge	
19. Art 100 & 225-6b(FPN)		20. C	21. C	22. B
23. 2..2..3	24. B	25. A	26. D	27. B
28. B	29. B	30. D	31. C	32. B
33. B	34. B			

Test #6:

1. Art. 100	2. A	3. C	4. B	5. A
6. C	7. A	8. D	9. A	10. B
11. D	12. Art 100	13. D	14. B	15. B
16. A	17. C	18. D	19. B	20. B
21. D	22. 0..0..0..0	23. See 310-4		

Test #7:

1. 112.8	2. 17 conductors	3. 2"	4. 22.8a	
5. #6 copper	6. 13	7. 39.5 ci	8. 3/4"	9. 3/4"
10. 180.4a	11. 42.64a	12. 211a	13. 24.4a	
14. 31.25 sq.ft.	15. #3/0	16. 2.14 sq.in.	17. #14 SO	18. #12cu

Sample Exams for Electrical Licensing......Michael G. Owen